Author: Tero Isokauppila
Artwork and layout: Juho Heinola

Printed in the United States by: Print is Art Team at GLS, Mpls. MN

ISBN: 978-0-692-17731-0

www.santasoldshrooms.com

Published by Four Sigma Foods, Inc.
1450 2nd St, Santa Monica, CA 90401
www.foursigmatic.com

© Four Sigmatic 2018

THE ORIGIN STORY OF THE
WORLD'S MOST FAMOUS PERSON

Tero
Isokauppila

FOUNDER OF

FOUR SIGMATIC®

Story Time With Daddy

O nce upon a time, there lived a curious young girl in a small village far away. The girl was so curious that her favorite thing to do was to listen to stories from the village elders. She would listen to them all day, and then rigorously fact-check their wild stories online in the evening.

She was only ten years old, but already she wanted to know everything about everything.

Late one winter night the young girl came to give her father a hug and kiss goodnight. It was bedtime, and this was their tradition, but on this night the young girl was not tired. Her curiosity had gotten the better of her.

"*Daddy, tell me a story,*" the young girl begged her father, as she jumped into his lap. Her father was not one of the village elders, but the girl thought he was the smartest person in the whole world.

"*Would you like to hear the real story of Santa Claus?*" her father said. Christmas was around the corner, so there was no time like the present, he thought.

"*Yes, very much,*" she replied as the fire crackled and three stockings—Daddy's, Mamma's, and hers—hung from the chimney.

"*Well, okay then.*" Her father turned off the television, and began.

CHAPTER 1

The Soda Saint

Y ou've seen pictures of Santa Claus on the side of a soda can before—yes, my darling?" her father asked. The young girl nodded. *"He is a jolly old man with a red velvet suit and white hair and a bushy white beard to match. He has a wonderful, magical spirit, doesn't he?"*

"He does, Daddy," the young girl replied.

"Well actually, Santa Claus is a modern term for Saint Nicholas," her father continued. *"'Santa' means saint, and 'Claus' is a German nickname for Nicholas, which is why so many people think Santa is German."*

The little girl looked up at her father as if she were expecting to hear more.

"But he's not German. St. Nick was a Turkish bishop, famous for secret gift giving to those in need. His legend spread to Western Europe where St. Nicholas later became Santa Claus."

The father looked down at his daughter, waiting for her to be impressed with his knowledge, the way she was with the village elders. She was enjoying the story, but this was not news to her.

"But Daddy, St. Nick is based on an older story about 'Old Man Frost!' He was a Slavic winter wizard with white hair, long beard, a fur-trimmed robe, and a magic staff. He kinda looks like Gandalf the White," the young girl said matter-of-factly

"Interesting," Daddy said, smiling. *"Do you think that's where the Santa image from the soda can came from?"*

"I do!" the young girl said.

The young girl was right. And it would not be the last time on this cold winter's night.

CHAPTER 2

More Like North Pole Adjacent

here was I?" her father continued. *"Oh yes, and for over 150 years, Santa has been said to live at the North Pole with Mrs. Claus. But we know that's not exactly true, because..."*

"...the North Pole is covered in shifting ice in the middle of the ocean," the young girl said.

"That's right," her father said, proud of his little darling's wisdom.

"That place is too cold, Daddy. Even for Santa," she said with confidence.

"Where do you suppose he lives then?" Daddy asked.

His little sponge clearly knew more than she let on when he began the story. Now it was time to see who knew more. These were some of his favorite moments with his daughter as she grew from a baby into a little person.

*"He lives just below the North Pole in a place called **Lapland**."*

Right again. Lapland is a region that spans throughout the northern-most parts of Finland, Sweden, Norway, and Russia. It's the closest habitable land to the North Pole.

"There are snowy mountains and peaceful valleys that glow with the Northern Lights." the little girl continued. *"It's like a real-life snow globe! Daddy, doesn't that sound like the kind of place Santa might live!?"*

"How do you know all this about Lapland?" her father asked. His little girl looked up into his eyes.

"Google Earth," she said.

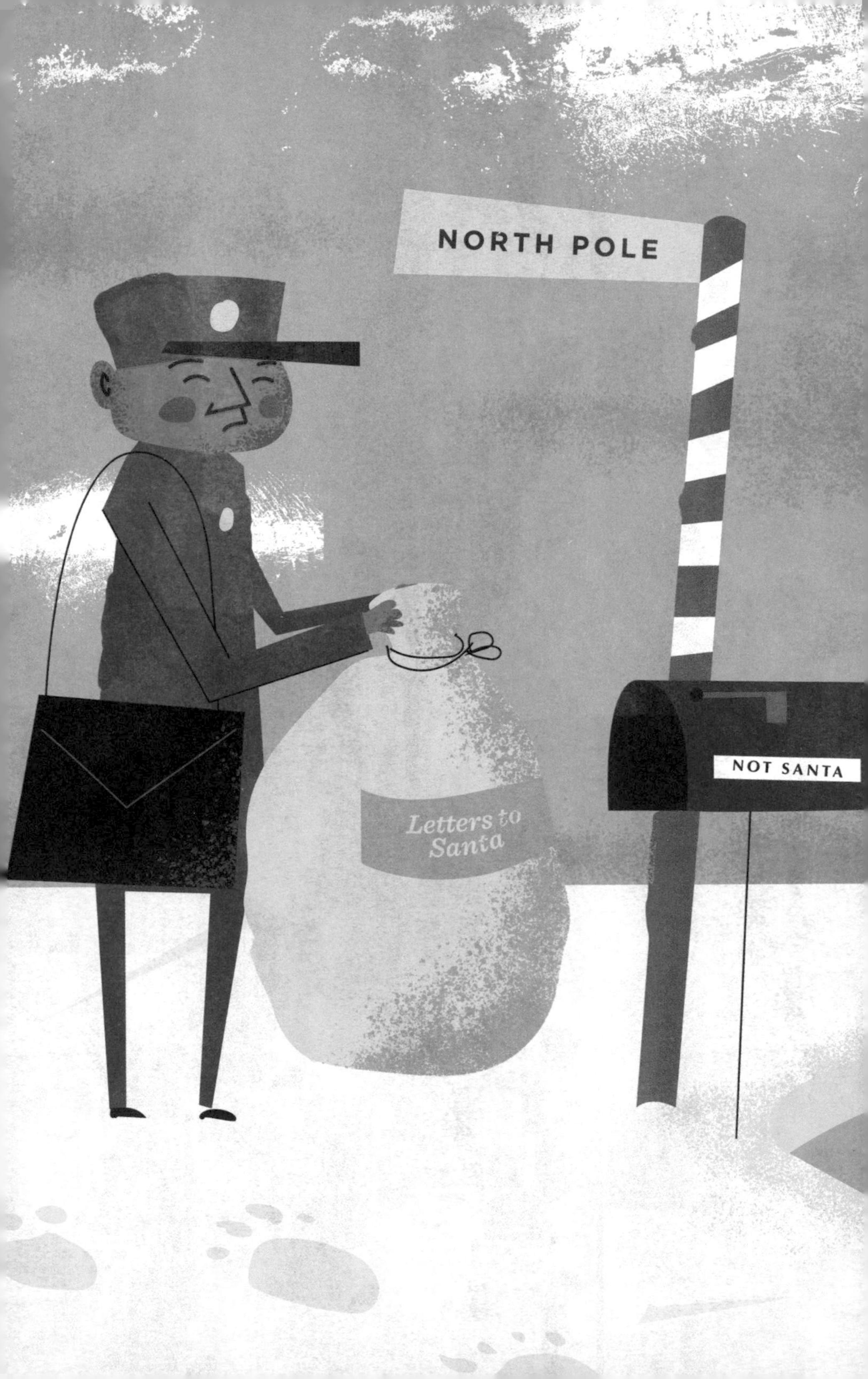
NORTH POLE
Letters to Santa
NOT SANTA

CHAPTER 3
Reindeer People, Who?

"Did you see the reindeer on the computer, too?" her father asked.

"*No!*" the little girl said excitedly.

"*I'm surprised,*" he said. Lapland is a home to one of the world's only population of native reindeer. "*What about the Reindeer People? Did you see any of them?*"

The young girl's eyes grew wide, as they did every time she learned something new.

"*Who?*" she asked.

"*They're called the* **Sami**. *They're nomadic reindeer herders,*" he explained.

"*Nuh-uh!*"

"*Yea-huh.*" Indeed, the Sami make everything they need to live from reindeer meat, bone, hide, and sinew. "*They're also known for their colorful, embroidered clothing.*"

"*Like my Anna and Elsa dolls!?*" the young girl said. This was getting to be almost too much.

"*Except they're real,*" he said.

"*Omigoodness this is so amazing,*" the young girl said, barely containing herself. Then suddenly all the features of her face contorted in a twist. Her father recognized this look. It was her "question face."

"*But wait, what do the Sami have to do with Santa?*" she asked.

"*Santa is their spiritual leader.*"

CHAPTER 4

Santa Was a Shaman

T he young girl's mind was spinning with possibilities. On the one hand, now it made sense why Santa was so good driving the team of reindeer at the front of his sleigh. But, how did being the spiritual leader of an unknown tribe make someone good at giving presents? Or get them to every shopping mall during a December shopping spree?

"The Sami believe that all things have a spirit," her father began, by way of explanation. *"Animals, plants, rocks, the wind—everything."*

"Okay," the young girl said, unsure of where this was going.

"To connect with the spirits, the Sami make offerings to nature during big ceremonies. One of the largest happens at the winter solstice. Can you think of another ceremony that involves offerings and takes place in the winter?" he asked.

"Christmas," she said. He knew she would know that. And if he kept going along this track, she would arrive at the answer to her own question. There was just the matter of this next bit.

The Sami responsible for performing these solstice ceremonies is the shaman. Equal parts medicine man and priest, he visits each family's home to conduct the ceremony. With a little help from nature's pharmacy, he becomes the bridge to the spirit underworld.

See, how do you explain this to a ten-year-old? It's not easy. So he tried to sidestep it.

"The Sami winter solstice ceremony is the ancestor of Christmas. In Lapland it eventually became known as Joulu. Then Joulu became Yule in German-speaking areas, which later turned into Christmas."

"But...how?" the young girl asked.

"Okay, picture this." Daddy said. *"One special, snowy night a reindeer-drawn sleigh pulls up to your doorstep. Upon it sits an older wizard-looking man wearing a fur-trimmed coat, boots, and a beard to warm him up in the biting cold. Over his shoulder is a heavy sack full of offerings."*

"That sounds like Old Man Frost!" the girl said, connecting the dots.

*"Bingo. That ancient Sami shaman was called **noaide**. But since then, he's had many names, from Old Man Frost to St. Nick to Father Christmas."*

CHAPTER 5

How Christmas Got Its Name

The idea of Santa Claus as an ancient Scandinavian medicine man had the young girl on pins and needles. Her father looked up at the clock. Now it was really past her bedtime.

"We've had enough for tonight," her father said.

"No, but Daddy, I don't understand," the young girl said. *"Why is it called Christmas then?"*

This was a good question. In Lapland, the winter solstice marked the most significant time of the year. It's the end of continuous darkness and the return of the light. The Sami took this as a time to reflect inward and prepare themselves for the year ahead. This tradition merged with the Yule celebrations centuries later. So why wasn't it called Noaide-mas or Winter Solstice Day?

"It's complicated, darling. When the Christians came along, they adapted the solstice celebration to honor Jesus's birth," the father explained. *"And I guess since it's all about new hope and embracing the return of the light, it seemed like a good fit to them."*

"So Santa isn't related to Jesus?" the young girl asked, her curious ten-year-old mind now fully in gear.

"Not really," her father began.

"That's not true at all," a voice came from behind them, cutting through the noise of the crackling fire. It was Mamma Brigitta, the young girl's grandmother, carrying in a tray of warm apple cider from the kitchen.

"They have a lot in common."

CHAPTER 6
Santa and Jesus

M amma Brigitta grew up in Lapland and was visiting her son and granddaughter for the upcoming holidays. She had become a regular visitor ever since the young girl's mother passed away when she was five. To the young girl, she was just Mamma, but her parents affectionately called Mamma Brigitta the "crazy, hippie grandma." And she was about to find out why.

"How exactly are Santa and Jesus connected again, Mom?" the young girl's father asked with a smirk. *"Is Santa an independent contractor working for Jesus? Do they trade notes on who is naughty or nice? Do they drink eggnog together? Is it spiked with wine?"*

Mamma Brigitta raised her boy on stories just like this one, but all his formal schooling made him cynical about her ancient wisdom. She knew what her son was doing with all his questions and she would not take the bait. Plus, it did not take a genius to see the connection. Santa and Jesus are some of the most recognizable figures in the world. Each is known for their supreme gifting powers. They're both into the whole "good" and "bad" thing. And everyone is always asking them for stuff.

"When you look at Santa's shamanistic roots, there are a lot of similarities," she said calmly.

"Do you think this is appropriate talk for a little girl before bed?" the young girl's father asked.

"Give her some credit, Sven." Mamma Brigitta said.

"Yeah, Daddy, I'm practically in middle school," the girl said proudly.

*"Besides, it's not like I brought up the **Amanita muscaria** or anything."*

"Mother!"

"What's the Amanita musc...musc...?" the young girl asked, both confused and curious.

"Amanita muscaria," Mamma Brigitta repeated. *"It's a very special mushroom."*

CHAPTER 7

The Lost Symbol of Christmas

T his was not how the young girl's father was expecting story time to go. And he knew he was about to lose control of it, because he knew about this super mushroom. It is, by most accounts, the world's most famous mushroom that nobody actually knows about. You've seen the red and white toadstool in old video games and most garden gnome statues. Ever used the mushroom emoji? Then you know this mushroom—by sight at least.

"What makes it so special?" the young girl asked, looking first at her father and then to her grandmother.

"Remember when I read Alice in Wonderland to you last time I came to visit? How Alice fell into a rabbit hole, ate pieces of the mushroom and started seeing things?" Mamma Brigitta asked.

"Oh, you mean magic mushrooms?" the young girl asked.

"You swore you'd keep this hippie stuff to yourself until she was in college," Sven raised his voice to his mother.

"Wasn't me," Mamma Brigitta said.

"How on earth do you know about that?!" the young girl's father asked.

"From the internet," the young girl said.

"Well, that's the internet for you. You can't have the sweet without the salt," Mamma Brigitta offered in defense of her granddaughter.

"What is it?" the young girl asked. *"What does it do?"*

"This is a ceremonial mushroom," Mamma Brigitta explained. *"The Sami are one of many indigenous peoples for whom it is a sacred, spiritual tool."*

"But what does that have to do with Christmas—or Jesus?" the young girl asked.

"Only everything," said Mamma.

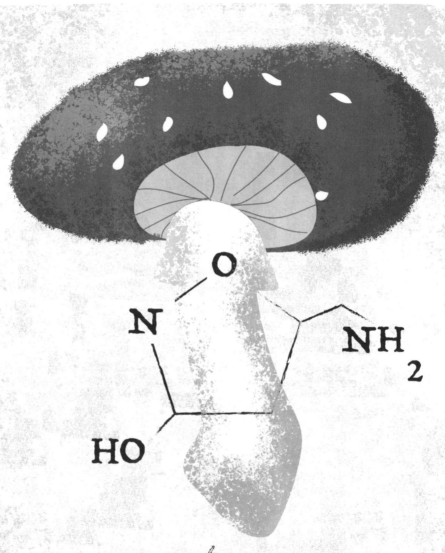

amarita muscaria

36

CHAPTER 8
The First Christmas Gift

Today, Christmas icons are credited to Jesus, but most of our Christmas customs are actually Yule traditions. Yule traditions come from 5,000-year-old Sami traditions. And the Sami traditions link back to this red-and-white magic mushroom.

"Guess where this Santa mushroom grows?" Mamma Brigitta asked her granddaughter.

"Lapland?" she said.

"Yes, but where exactly?"

"I don't know," the young girl admitted, looking to her father for a hint.

"They grow under spruces, pines, and firs," Daddy said.

"Christmas trees!" the young girl exclaimed. Her eyes lit up. She loved when new bits of knowledge started to make sense with what she'd already learned. Her face only got brighter as her grandmother described how the mushrooms are ripe for the picking around late summer. And that the shamans then need to dry and preserve them for the winter solstice ceremony.

"Shamans would hang the mushrooms on the branches of large spruce trees to dry," Mamma Brigitta explained, conveniently leaving out the part where the drying process intensifies the psychoactive properties of the mushroom. *"Just imagine: a large tree covered with red-capped mushrooms symbolizing the coming holy day."*

The father directed his daughter's attention to their own Christmas tree in the corner. It was a large spruce tree covered in tinsel, lights, and a handful of decorations. And, shining through it all, dozens of red glass baubles spread across the ends of all the big branches.

"And guess what the shamans did with all those drying mushrooms once they were done foraging?" Mamma Brigitta said, recognizing that she had her granddaughter in thrall. *"They brought them home and finished drying them by putting them in socks over the fireplace."*

The young girl looked at her Christmas tree, and then at the mantle of her fireplace. It was like she was staring into the living embodiment of the real Christmas past. This was the most amazing thing she'd ever seen or heard in her life. "This is so cool," she thought to herself. How had she not found it anywhere online yet? "A lot of good you are, dark web!"

CHAPTER 9

High as a Kite...or a Reindeer

S o how does the solstice ceremony go?" the young girl asked. *"With the mushroom, I mean?"*

"Every year the shaman spent the days leading up to the actual winter solstice going from hut to hut to perform his special ceremony for each family."

"That's a lotta ceremonies, Mamma. Doesn't he get tired?" the young girl asked.

"I'm sure he did, sweetheart, but he also had a strong team of reindeer pulling him in his sleigh, so that helped."

"Being able to fly from house to house saved lotsa time, huh?" the young girl said sweetly.

"Reindeer can't fly," her father said, *"you know that, right honey?"*

"In a manner of speaking, that's not exactly true," Mamma Brigitta corrected her son.

"Mom, I thought we agreed. No ippie-hay, uff-stay."

He was forgetting a part of the story his mother had told him dozens of times as a child. The reindeer of Lapland had a very unique diet. They loved Amanitas. In fact, it was their rooting around under the spruce trees that drew shamans to Amanitas in the first place. Like catnip does to cats, reindeer go nuts for them. Upon eating them, they jump and leap around excitedly, kind of like they're flying.

Maybe it was age, or the responsibilities of fatherhood that allowed this to fade from her son's memory, Mamma Brigitta suspected. She wasn't going to let that happen to her granddaughter.

"It's not just the reindeer, either," Mamma Brigitta said. *"One of the most common experiences of people who eat Amanitas is the sensation that they, too, are flying."*

"Is that what they mean by 'high as a kite' on all those dentist viral videos?" the young girl asked her father, looking up at him, waiting for his eyes to give away the answer, which they did. He smiled, holding back a laugh, as the young girl giggled to her Mamma.

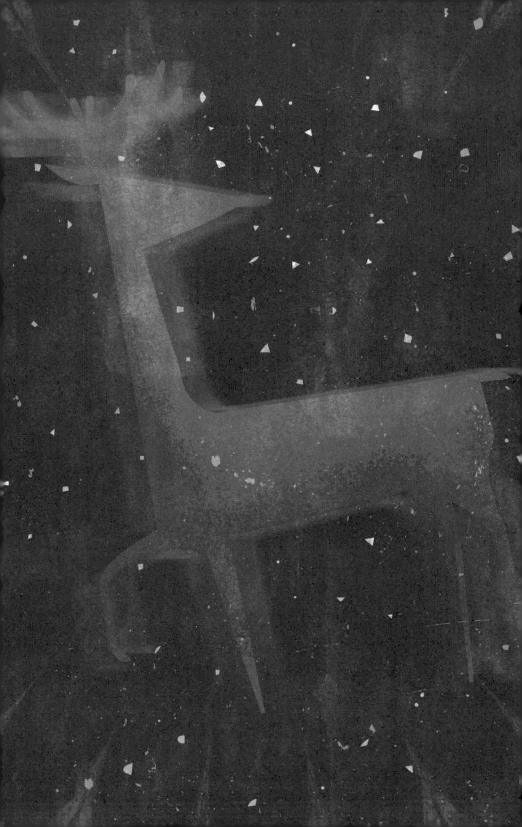

CHAPTER 10
In Through the Chimney

T he young girl's giggles quickly turned into a big yawn. Her father checked the clock again. She was now officially "up late." And while her eyes were heavy, her mind was still light and nimble.

In her mind's eye, the young girl could practically see the Sami shaman. Wearing a furry reindeer robe, carrying a sack of shrooms, he arrives on a sleigh drawn by a "flying" Rudolph. It's basically Santa.

Just then, a new log on the fire snapped, crackled, and popped, sending a plume of smoke up into the chimney. Once more the young girl's expression twisted into her "question face."

"Daddy, Santa doesn't really come down the chimney, does he?" she said, more statement than question. It was the first time one of her questions felt tinged with disappointment.

"Actually, my darling, the Sami shaman did come in down the chimney," her father said.

"What?! Really?!! Are you messing with me, Daddy?" she said, desperate for this to be true.

"Mom, would you like to tell her?" he said to Mamma Brigitta.

"You know the story, Sven," she replied, her beaming smile lighting up her face.

"The Sami people lived in a kind of teepee made from cloth, wood, and moss called a **kota**,*"* he told his daughter. *"And at solstice time, in the dead of winter, the doors of the kotas were usually snowed over. This was Lapland, remember. It's..."*

"...near the Arctic Circle," the young girl broke in excitedly. *"Brrr."*

Her father smiled and continued: *"Every kota had a fireplace to keep the family warm. To let the smoke escape, there was a hole at the top, like a chimney. And if the kota got snowed in, that was the only entrance or exit. The Sami would use a ladder to climb in and out whenever necessary, avoiding the blazing fire below. So at solstice time, when the Sami had guests, like the shaman, they would often come in through this 'chimney.'*

The young girl could not have been more excited. She'd always felt like the story of Santa she'd heard on TV wasn't totally true, but the image of Santa coming down the chimney had always been one of her favorite parts of the story.

CHAPTER 11

It's Beginning to Look a Lot

Like Solstice

The young girl nestled deep into her Daddy's chest, warmed by the fire and comforted by the knowledge that Santa did come down the chimney. Mamma Brigitta then picked up the story of the solstice ceremony where she had left off.

"Here's where the real magic begins," she teased. *"From his sleigh, the Sami Santa grabs his sack full of gifts and enters through the chimney. Inside the kota, with the fireplace roaring, he eats a healthy dose of Amanitas and beats a ceremonial drum while he chants."*

"And from all that heat and chanting and drumming," her father said, *"he starts to turn bright red."*

"That's why Santa has rosy cheeks, not because he has bad skin!" the young girl shouted with glee.

"It's one of the classic effects of the Amanita," Mamma Brigitta said. *"And once it kicks in, the shaman goes into a trance and starts to have visions."*

"Does it hurt him, Mamma?" the young girl asked.

"It's not always rainbows and reindeer," Mamma Brigitta admitted. *"Sometimes he might become sick to his stomach, or become weak and go into despair. Other times, thanks to a chemical in the mushroom called muscinol, he falls over and giggles like crazy."*

"Despair?" the young girl said. For a ten-year-old, it's an easy word to say but not so easy to understand. Neither adult wanted to get too deep into that one this late at night.

"Mushroom visions can include nightmares, speaking with the dead, or other intense experiences like that," Mamma Brigitta said.

"Scary," the young girl said.

"It can be," father Sven said, *"The Sami say that at this point the shaman will release his spirit from his physical body."*

"But, but, where does it go?" she asked.

CHAPTER 12

The World Tree and the North Star

Once the shaman's spirit leaves his body, it's free to travel anywhere in the skies, the oceans, or the underworld. On its travels, the shaman will ask the spirits and dead relatives if they have messages for the family. This was a hard concept to explain to the young girl. Heck, it's a difficult subject to wrap your mind around as an adult.

"But Mamma, how does his spirit get up to the sky or to the underworld and stuff? Does it fly?"

"It climbs the World Tree," Mamma Brigitta said.

"What is THAT?!?" the young girl asked, riveted by this new concept.

The World Tree is not a new idea. Many ancient peoples believe in a World Tree—a cosmic, holy axis and origin of the universe. The World Tree's roots stretch down into the underworld, and the branches reach the heavens. For the Sami, the world tree even reaches the North Star, Polaris, which is the brightest star you can see from Lapland. In the solstice ceremonies, the shaman's spirit would climb the world tree and touch the star. It represented truth and wisdom, which is why we now put a star on the top of the Christmas tree.

"For the Sami," Mamma Brigitta explained, *"Pinaceae trees were holy in that same way."*

"Those are the Christmas trees!" the young girl recalled.

"Uh-huh, very good," her father said.

"And the Santa mushroom can't grow without them—the spruce or the fir or the pines."

"Daddy, remember how I told you about my science class last week?" the young girl said. *"Our teacher said that the pineal gland in our brains is named after the pine cone."*

"And guess what," Mamma Brigitta said teasingly, her granddaughter's eyes brightening once more. *"That gland is the center of our human soul."*

"My soul is in my brain?"

"In your mind," Mamma Brigitta said. *"When the Sami talked about the mushroom tapping into the god within, that is what they meant."*

CHAPTER 13

Okay, Santa *Traded* Shrooms

A s mature as the young girl was for her age, she didn't yet have the words to describe all this talk of soul glands made out of pine cones. When she gets older she'll be able to recognize that this part of the story was *"metaphysical AF."* In the meantime, though, it brought her back to more basic questions.

"How long does it take for the shaman's spirit to climb the World Tree?" the young girl asked.

"It can last several hours," Mamma Brigitta said.

"What happens when he gets to the top?"

"The shaman returns to his body, completely exhausted. The family in the kota waits for him to gather up his energy so he can relay the messages he received from the spirits."

"What kind of messages?" the young girl asked.

"Wisdom from their ancestors. Visions for the New Year. Other revelations from beyond," Mamma Brigitta explained.

"Neat," the young girl said.

"Based on his visions, the family will make offerings to restore balance in their life. They will also make payment to the shaman with gifts of goods, food, especially reindeer milk."

"It's too cold for cows in Lapland, I bet," the young girl said. *"M-m-m-m-ooo-ooo-ooo."*

"Know who else was waiting for a gift?" Mamma said through a chuckle. *"The reindeer."*

"Mooooommmm!" said her son, suddenly 10 again, himself.

CHAPTER 14

~~Don't~~ Eat the Yellow Snow

O utside the kota, the shaman's reindeer are waiting for him to come out and relieve himself on the snow. Don't worry, it's not because they're perverted and want to watch. Nothing gross like that. They just want to eat it.

What's that now? Why would any animal ever lust for a pee-flavored snow cone? Well, the magic parts of the Amanita are not all metabolized by the body and remain active in the urine. In fact, the shrooms remain potent after six passes through the human body. It's actually safer and more powerful to drink Amanita urine than to eat the mushrooms whole.

This is why drinking urine was not rare among indigenous cultures. The shaman may even drink his own urine to intensify the effects of his trance. It was also how those in attendance could join the shaman on his journey into the underworld.

Now, try explaining that to an inquisitive ten-year-old. The young girl's father didn't even give it a shot. He just laughed and asked his daughter: *"Do you remember how your British cousin Hannah said she 'gets pissed' at her college parties?"*

"Uh-huh," she said through a big yawn, fighting off sleep finally.

"That expression came from the solstice ceremony. That's all you need to know."

CHAPTER 15

Naughty and Nice

ith one final, sweet yawn, the young girl was about to fall asleep. Mamma Brigitta kissed her on the forehead. Her father scooped her up from his big, cozy chair by the fire.

"Did we ever tell you the real story of the Christmas elves?" her Daddy whispered in her ear as he ascended the front stairs up to her bedroom.

"Nnnn," the young girl muttered in a barely audible voice that only her father could decipher. He knew that she was about to fall asleep, but he couldn't pass up a special moment like this to tell her one of his favorite parts of the Christmas story. Who knows, maybe some of it would sink in.

"In Scandinavia," her father whispered, *"there are these tiny beings called* **tomten**. *They live in people's barns or saunas and watch over their family. They look like little garden gnomes—small with a long, white beard and colorful, pointed cap."*

Her father looked down at his daughter as he opened her bedroom door. She was almost asleep.

"Sure, they bring good luck, but the tomten were a mischievous bunch, like leprechauns. When you think nobody is looking, the tomten are. They watch your every move both day and night to see if you are being bad or good. The Finnish call them Joulutonttu, Christmas elves! How do you think Santa watches that many of you rascals at once? With an army of tiny gnomes."

Her father tucked the young girl into her bed.

"So you better watch out. You better not cry," he said smiling, *"because if you're bad, a tomten will tell Santa and make you very unlucky, indeed."* He kissed her on the forehead. *"If you're good though, they'll leave you generous gifts at your front door once a year at Yuletide. So leave the tomten a thank you bowl of porridge on the doorstep—and don't forget the butter, or he'll play a nasty trick on you."* Her father paused to admire his precious daughter one last time. *"You can also leave a glass of reindeer milk for Santa,"* he said with a chuckle as he backed slowly out of her room and turned off the light.

CHAPTER 16
The Son of God was a Fungus

T his is not how I thought tonight's bedtime story would go," Sven said to himself as he headed back downstairs. His mind began to drift from his daughter to Santa, then to Jesus and, of course, mushrooms. It was the curse of a hippie mom, he figured.

He thought back to his college days and one of his favorite books from that time. It argued Jesus was not a man, but rather—are you ready—a mushroom. One that, over the centuries, had become anthropomorphized, or given human attributes. While he never took this argument literally, he started to believe that Jesus was on shrooms.

Mamma Brigitta was still in the living room when her son returned. A smirk filled her face.

"What's so funny?" he asked, feeling defensive.

"You. Me. Us," Mama Brigitta said. *"How we humans love to take sacred rituals and scrub them of the details that make us uncomfortable. Even you, whom I have told this story many times, had forgotten bits and pieces."*

"It's just one story, mom," he replied.

"...But is it?" she continued. *"Remember that book you read in college about Jesus being a mushroom? The one you couldn't stop talking about. Where did that curious boy go?"*

His mother always did this. She'd mention something just as he was thinking about it, like she could read his mind. When he was younger, it drove him crazy. He actually thought she was telepathic, which she never fully denied, just as you would expect from a hippie mother.

"Mushroom iconography shows up again and again in Christian art," Mamma Brigitta continued. *"The halos behind Jesus and the saints definitely resemble the mushroom cap. The Holy Grail could easily be an upside down Amanita cap. Or the literal sacrament of Christ—the symbolic blood and flesh served at The Last Supper."*

"Yes! That's how Jesus turned water into wine—Amanita urine." her son said snarkily.

"Don't get smart with me," Mamma Brigitta said laughingly.

Whatever the ultimate truth may be, the man couldn't deny that Jesus's name has been linked with mushrooms for two thousand years. And that has to mean something.

CHAPTER 17

So, Santa Was a Spiritual-

Shaman-Healer. What Do I Care?

There is a name for this process of sanitizing ancient wisdom for modern consumption. It's called spiritual bleaching. And the worst part of it, at least as far as Mamma Brigitta was concerned, was our tendency to repackage it for sale.

"There are many examples where we've taken sacred indigenous rituals and turned them into businesses," Mamma Brigitta said. *"Just look at your own daily habits."*

"What did I do?" her son said. *"I'm not the enemy here. I agree with you!"*

"For example, the sauna in your garage was used on days-long, life-changing vision quests." Mamma Brigitta said, unmoved by his protest. *"Today, they are in almost every gym as a pleasure carrot at the end of a workout.*

You also like to practice yoga, which is an Indian path to enlightenment. Today, you can find 'hot power sculpt' classes on every block. Yoga has become an American path to a quick high and washboard abs."

"Ok fine, but are you any better, mom?" the son countered. *"I saw the profile you made on my computer for the new senior singles dating app. 'My name is Brigitta and I am fascinated by human nature because I'm a Virgo.' You used the spiritual bedrock of every ancient culture to get yourself a date!*

That's why I liked when we were kids how you always told us the story behind every holiday," Sven said. *"Otherwise I would have ended up mainlining margaritas and eating chips and guacamole out of a sombrero on Cinco de Mayo, yelling 'Happy Mexican Independence Day', even though that's not what it is."*

*"And that's **spiritual cleansing** at best: you forget the old traditions and lose the real stories. You might even completely dismiss the dark past of holidays like Thanksgiving."* finished Mamma Brigitta.

That is exactly what happened to the Sami. The church and the crown targeted their shamans in the 19th century. They accused them of heresy and witchcraft. And like most indigenous cultures, the Sami had no choice but to submit to the stronger powers. As a result, the Sami's sacred indigenous traditions slowly disappeared. Eventually the age of shamans came to an end, while Jesus and St. Nick rose up in their place. Talk about winning the game of thrones.

CHAPTER 18

A Drug by Any Other Name

Mamma Brigitta and Sven agreed that cultures can also be wiped out by the removal of their sacred customs, not just through death and war. History is written by those victors. And for now, we are them, with eraser in one hand and permanent marker in the other.

Yet, mother and son still differed on what the young girl should know now about this story. Sitting quietly in the living room, sipping cider, the issue was still weighing on Sven's mind.

"Do you think it was smart to bring up the magic mushrooms?" Sven wondered aloud, finally. *"Your grandchild is very smart and could easily find these mushrooms online."*

The most common magic mushroom, psilocybin, is known for a euphoric, one-with-the-universe trip. The Santa mushroom, on the other hand, is a dark psychedelic that, while not lethal, can hurt you. It operates in the body the same way as alcohol and popular sleep drugs. One or two can cause nausea and distorted sight and sound. Six will make you ill and knock you on your bottom.

Sven knew that the little red toadstool is related to two of the world's most poisonous mushrooms—the Destroying Angel and the Death Cap. In addition, Amanita muscaria's main compound (ibotenic acid) technically "eats your brain," so if she stumbled upon the wrong variety or had too much, it could be bad news. This was worrisome to the father of a curious child.

Mamma Brigitta was less concerned. *"Well, you know how I love my psychedelics."*

"Mom!"

"What? Our government has sold us soda with cocaine in it, cough syrup with heroin in it. Shrooms are way less dangerous than those," she started to preach. *"They can even help people with trauma and terminal illnesses. Look at the research people are doing! It's amazing."*

"Yes mom, I agree with you. And one of these days she will need to understand that," her son replied. *"That the war on drugs is politics. That not all intoxicants are equally bad, and some are actually good. But that day is not today."*

"You're right, you're right. I forget how young she is sometimes. She has that same curious mind you had," Mamma Brigitta replied. *"I'll talk with her tomorrow."*

"Why don't I do the talking this time? I know what to say. You taught me well," he said.

CHAPTER 19

'Tis The Season

S ven and his mother had not connected this deeply since the death of his wife. The young girl's curiosity tonight brought back a connection that had been lost for a long time.

It was only fitting that the peculiar origin story of Santa Claus made this happen. After all, the main purpose of the Sami tradition was not to eat shrooms and see monsters, but to connect families together year after year. Indeed Christmas was, and is, a celebration of generosity, gratitude, and family time. This is the real magic of the winter solstice—not its materialistic sugar coatings.

Inspired by the Sami, we too can celebrate Christmas with joy and music. Light a candle for loved ones lost. Meditate on our deep connection to the universe or God. Honor Mother Earth and make offerings to her. Go on a spiritual journey or retreat. Seek guidance from community elders. Invite messages from the spirit world. Use intoxication as a means to connect, rather than as a life vest, like so many of us do, to get through the holy-days with our disconnected families.

All of this can help us enter the upcoming year with increased awareness and good spirit.

Perhaps in this way, the **"Christmas magic"** will radiate into our lives.

Because it's real.

Shifting to fact from fiction

A re all the facts in this book 100% true?
Probably not.

But there are plenty of legitimate sources verifying Santa's shamanistic roots. Still, like any legend that has endured both time and travel, Santa is a mysterious blend of fact and fantasy.

The jolly old man has likely drawn influence from multiple Pagan stories. From the Italian grandmother Befana, who filled children's stockings with gifts, to Thor's goat chariots riding into Asgard (kind of like Santa rides that reindeer sleigh), or Poseidon's habit of galloping across the sky each winter solstice, there are many theories about the origin of everyone's bearded buddy. Like some of the best cuisines today, the Christmas story has been blended with many local favorites.

But, there are two things of which we can be sure:

1) CHRISTMAS IS ACTUALLY A PRE-CHRISTIAN HOLIDAY WITH MYSTIC ROOTS IN SHAMANISM.

2) AN AWESOME MUSHROOM PLAYS AN UNEXPECTED, KEY ROLE IN THE BELOVED TALE.

Part psychedelic and part poisonous, the Amanita mushroom changed the course of gifting forever. And it may have had an influence on Christianity, too. One day the Sami shaman could even surpass Jesus as the global poster boy for Christmas.

Perhaps we should no longer laugh at kids who believe in Santa and elves.

After all, if Santa Claus is real, what other everyday magic are we missing out on?

About Tero Isokauppila

Tero is the founder of *Four Sigmatic*, a nature-centric Finnish-American company specializing in functional mushrooms, superfoods, and adaptogens.

Tero's roots (or mycelium, if you will) are in Finland, where he grew up foraging for mushrooms and other wild foods on his family's farm. He later earned degrees in *Chemistry and Business*, as well as a *Certificate in Plant-Based Nutrition at Cornell University*. In 2012, Tero founded Four Sigmatic with the dream of bringing a little Everyday Magic™ to the lives of all.

An expert in all things related to mushrooms and natural health, Tero is the author of two books: *Healing Mushrooms: A Practical and Culinary Guide to Using Adaptogenic Mushrooms for Whole Body Health* (his first book from Avery Publishing); and this book, *Santa Sold Shrooms*.

Tero was chosen as one of the world's Top 50 Food Activists by the *Academy of Culinary Nutrition* and has appeared in *Time, Forbes, BuzzFeed, Vogue, Playboy, Harper's Bazaar,* and *Bon Appétit*. He is also a sought-after speaker, having been featured at *Summit Series, Wanderlust, WME-IMG,* and the *Fast Company Innovation Festival*.

In 2012, Tero left cold European winters behind in favor of sunny Southern California.

A lesser-known fact about Tero is that his middle name, Tapio originates from the Finnish god of forests, animals, and mystics.

About Juho Heinola

Juho (pronounced like U-Haul, just drop the L) Heinola is also a native of Finland. Juho has been an artist and designer for more than 15 years. He started out at the age of 11 as a comic book artist at the local newspaper and still has his own weekly comic in *Aamulehti*, which has been the second biggest newspaper in Finland for more than 10 years. Today, Juho has a dual role as an Art Director for *Four Sigmatic* and as the co-founder of *Creative Agency Snou*.

Juho has previously authored two books in Finnish; a children's book entitled *Mr. Zanzibar's Zoo is Moving* and a guidebook for every Nintendo Entertainment System game ever made called the *NES Atlas - A Guide for Nintendo Collecting*.

Besides Finland, Juho has lived in Israel, France, and Vietnam. He studied at the *Aalto University School of Art and Design* and at *Parsons Paris*. His package design work with Four Sigmatic was chosen as one of the best graphic designs in the United States in 2016 by *Graphic Design USA*.

Juho lives in Helsinki, Finland with his designer wife, Shani and beautiful daughter, Uma. His all-time favorite Christmas tradition is to play a video game where an Italian plumber tries to collect functional mushrooms to gain super powers to better fight monsters.

ACKNOWLEDGEMENTS

Tero Isokauppila

I wish to thank the following people for helping to make this magical book a reality.

Pirkko, Markku, Vesa, Elina, and Antti for providing me with a loving and supporting family. Thank you for being understanding of my unique path in life.

All my fellow "shroomates" at Four Sigmatic. Without all of you the Mushroom Mission would not be where it is today. Besides the professional support, I appreciate the friendship we've created. Special thanks to Lucy Sunday for helping with the creation of this book.

Juho Heinola for your artistic vision and design skills. Without your help this book would be much less magical.

Kelly MacLean and Nils Parker for helping me with the writing process. Without you two this book would not have turned out this good, nor would I have been able to finish it on time.

Jon Bier, Eric Jackman, Ashley Stahl, Mary Shenouda, Michaela Blaney, Sebastian Laine, Elizabeth Jarrard, and Max Lugavere for your feedback on the book drafts. You all helped me polish the final version.

Finland for raising me and teaching me about tonttu, joulu, and punakärpässieni.

Sami people for inspiring parts of this story.

There are so many amazing human beings that deserve credit for inspiring and supporting me, but to save your time I will thank them in person.

Finally, I would like to also thank all the fungi in the world. You're smart. You're loyal. You're full of wonderful surprises. I appreciate that.

Juho Heinola

I'm honored to thank my wife and daughter and my whole family. You are everything I could have ever asked for.

I'm grateful for all my friends at Four Sigmatic. It's been a wild and memorable ride, and I'm proud to be part of something this unique and special.

I would like to thank everyone who is making it possible for me to be creative and inspired; you have allowed me to do this for a living for most of my life. Thank you!